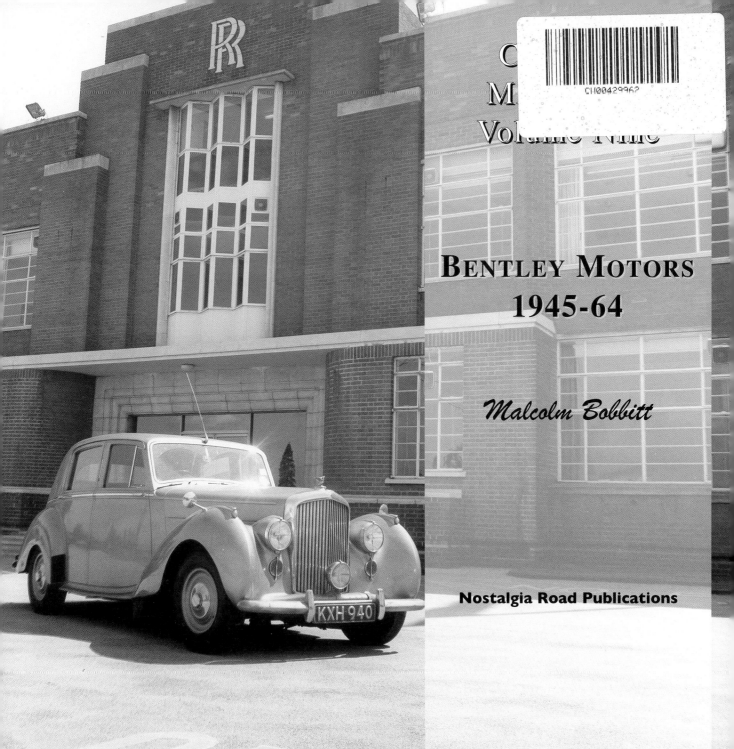

C[...]
M[...]
Vol[...]e N[...]e

BENTLEY MOTORS
1945-64

Malcolm Bobbitt

Nostalgia Road Publications

CONTENTS

The **Nostalgia Road** Series ™

is produced under licence by

Nostalgia Road Publications Ltd.
Units 5 - 8, Chancel Place, Shap Road Industrial Estate,
Kendal, Cumbria, LA9 6NZ
Tel. +44 (0)1539 738832 - Fax: +44 (0)1539 730075

designed and published by
Trans-Pennine Publishing Ltd.
PO Box 10, Appleby-in-Westmorland, Cumbria, CA16 6FA
Tel. +44 (0)17683 51053 Fax. +44 (0)17683 53558
e-mail: admin@transpenninepublishing.co.uk

and printed by
Kent Valley Colour Printers Ltd.
Kendal, Cumbria: Tel. +44 (0)1539 741344

Front Cover: *Continentals are among the most sought after Bentleys, this S2 having exquisite coachwork by H.J.Mulliner & Co. Beneath the bonnet lies a beautifully engineered V8 engine, which nearly 50 years on powers the current Bentley Arnage.*
Photographer: Malcolm Bobbitt

Rear Cover Top: *The Vilhelm Koren styled S3 Continental with coachwork by Park Ward was among the last Bentley cars to employ a separate chassis. The attractive four headlamp styling was well received by customers.*
Photographer: Malcolm Bobbitt

Rear Cover Bottom: *The MkVI was the first of the post-war Bentleys, the majority of saloons having Pressed Steel's 'standard steel' coachwork.*
Photographer: Malcolm Bobbitt

Title Page: *Bentley MkVI pictured at Crewe, the post-war home of Rolls-Royce and Bentley Motor Cars. The factory opened in 1938 to build Rolls-Royce Merlin aero engines and after the war was converted to car production, the first Bentleys leaving the works in 1946. The car depicted is fitted with Pressed Steel's standard steel coachwork and was photographed in 1949.*
Photographer: Rolls-Royce.

This Page: *Two generations of post-war Bentleys are seen here, the cars on the left being MkIV/R-types and that on the right an S3. All three models were built at Crewe, in a factory which was home to Rolls-Royce.*
Photographer: Malcolm Bobbitt

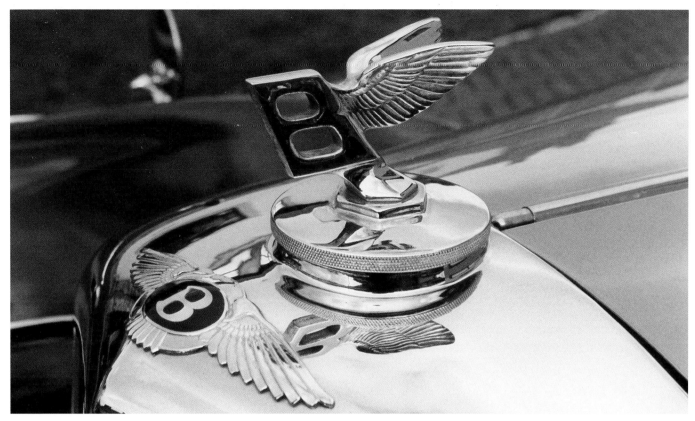

INTRODUCTION

Although the Nostalgia Road series has covered a wide range of transport topics - from refuse trucks to ice cream vans, it has never yet looked at a prestige car-maker until now and what could make a better starting point than Bentley? With its distinctive 'Winged B' configuration the Bentley emblem is one of the most revered to adorn any motorcar. Designed by F. Gordon Crosby, the emblem symbolises the marque's engineering and sporting heritage. Following Bentley Motors' establishment in 1919, the ensuing years were often turbulent for the company: in 1926 the firm was acquired by millionaire racing driver Woolf Barnato and in 1931 was absorbed into Rolls-Royce. Today, Bentley's custodian is Volkswagen.

The era of Bentley's fortunes discussed in this book is that between 1945 and 1964. These dates are significant, for it was in 1946 that Bentley began manufacturing at Crewe, in a factory built in 1938 to build Rolls-Royce Merlin aero engines.

Above: *The 'Winged B' emblem adorning the radiator shell, and the 'Flying B' on the radiator cap, famously depict this car as being a Bentley, in this instance a MkVI. The Bentley insignia was the work of F. Gordon Crosby who produced much of* The Autocar's *emotive artwork.*

The first Crewe Bentleys owed their origins to models that had been conceived pre-war. However, it was not long before design and development engineers were thinking ahead, and the highly respected MkVI and R-type models were replaced in the mid-1950s by the S-series Bentley, a superb design penned by John Blatchley who was Crewe's Chief Stylist.

Included in the Bentley catalogue was the exquisite Bentley Continental which, when introduced in 1952, was the fastest and most expensive British production car. It was then, and remains thus, a most desirable vehicle.

Malcolm Bobbitt   Cockermouth, August 2005

FOLLOWING A TRADITION

Crewe has been the home of Bentley since 1946, before which, the cars bearing the 'Winged B' monogram had been built at Derby, alongside Rolls-Royce motorcars. Bentley, however, was not always synonymous with Rolls-Royce: between 1921 and 1931 cars were built at Cricklewood in north London, where Bentley Motors relocated from a small two-storey workshop in New Street Mews off London's Baker Street. The cars built before 1931 are referred to as 'WO' Bentleys after their creator, Walter Owen Bentley (universally known as 'WO'), who was born in London in 1888. His creations occupy a legendary place in motoring history. For instance, they are renowned for winning the Le Mans 24-hour race on five occasions, initially in 1924.

Above: *W.O. Bentley at the wheel of his experimental Three-litre Bentley in 1920, the mud-splattered condition of the car implying that the engineer had been putting the vehicle through its paces ahead of being road-tested by Sammy Davis on behalf of* The Autocar. LAT

They then went on to win this famous endurance race four years in succession between 1927 and 1930. However, as famous as the cars frequenting Le Mans were the drivers, a brigade of wealthy young men whose ruthless exploits on the track matched their gregarious social lives away from it. They were known as 'The Bentley Boys', and their number included Sammy Davis, Woolf Barnato, Dudley Benjafield, Frank Clement and Sir Henry 'Tim' Birkin.

4

Top Right: *W.O. Bentley at the wheel of his DFP, with his fiancée Léonie Gore beside him, while waiting to attempt the Aston Clinton hill climb in June 1912. A novice at motor racing, Bentley was placed 'first in class' and went on to compete in the Isle of Man Tourist Trophy race.* LAT

Middle Right: *Bentley with Frank Clement (left) and Captain John Duff at Le Mans in 1924. Bentley was initially sceptical about entering long-distance endurance events but changed his mind when his Three-litre was victorious.* LAT

Bottom Right: *'WO' Bentleys have long been revered by enthusiasts, this example dating from September 1922 and photographed in 2004 in France. The car is extensively used by its owner, just as its creator intended.*

Bentley was educated at Clifton and though neither an academic nor sportsman, he was never happier than when dismantling a bicycle or other pieces of machinery that he reassembled to perfection. WO had a quiet but determined disposition: he was compassionate but nevertheless could at times be difficult, a particular look from him revealing far more than words alone! Rather than persevere with his studies at Clifton, he chose an apprenticeship with the Great Northern Railway at their massive Doncaster works. Whilst there, WO took a liking to motorcycles; he bought an ancient and well-used Quadrant which, after a year or so, was traded in for a Rex with which he began participating in competitive events. An Indian motorcycle replaced the Rex, and with it he entered the Isle of Man Tourist Trophy.

Having completed his apprenticeship, Bentley's aim was to become a locomotive designer, but this meant many years spent undertaking menial tasks, which he considered abhorrent. Forsaking railways for the motor industry, he joined a London taxicab operator to oversee the maintenance and daily running of vehicles. It was during this time that motorcycles gave way to motorcars, first a Riley and then a couple of Sizaire-Naudins.

When he tired of working with taxicabs, Bentley, in association with his brother Horace (HM), acquired the concession to sell French DFP motorcars. Thus Bentley & Bentley was established in March 1912: the brothers took over premises at Hanover Court Yard just off London's Regent Street, additionally attaining a small two-storey workshop in New Street Mews, off Baker Street. Whilst HM looked after sales and administration, WO undertook routine servicing and repairs, and preparing cars prior to delivery.

Above: *W.O. Bentley is pictured alongside the cars with which he became most associated. On the left is a 4¹/2-litre Bentley, the other a DFP. WO acquired the DFP franchise in March 1912, and it is this car that helped inspire him to create the Bentley marque seven years later.* LAT

WO believed the best way to promote DFPs was for them to be seen competing at motor sport events. A staid family machine regarded for its reliable performance, the DFP wasn't the typical sports car. Yet WO's engineering expertise changed that image when he won the Aston Clinton hill climb in June 1912, in a 'works' car tuned by himself. As other successes followed, WO driving his DFP, became a familiar sight at Brooklands and other venues, the pinnacle of achievement being placed sixth out of 21 starters in the 1914 Isle of Man TT event. It was through WO's efforts that DFP cars earned a distinct reputation for reliability and performance, largely because he was instrumental in using aluminium pistons. As an alloy, aluminium had been little used in engine design: today it is used exclusively, and for this we have to thank W.O. Bentley. WO shared his knowledge of aluminium with The Admiralty to further the war effort in 1914, the outcome being that the allies' air power was considerably enhanced. Having been sent to Gwynnes at Chiswick to help resolve problems manufacturing the French Clerget aero engine, WO was then dispatched to Rolls-Royce at Derby.

It was as a result of WO's discussions with Ernest Hives, who was head of Rolls-Royce's experimental department, that the Eagle, the firm's first aero engine, was fitted with lightweight aluminium pistons. W.O. Bentley went on to design and build his own rotary aero engines, the BR1 and BR2, which were used in a number of aircraft, including the Sopwith Camel, the front-line fighter aeroplane of the Royal Flying Corps (later RAF).

Top Right: *Pictured at Le Mans in 1929, when Bentley won the 24-hour race for the fourth time, Jean Chassagne is shown taking over from Frank Clement, the 4^1/$_2$-litre wearing no.8 and being placed fourth at the end of the race.*

Middle Right: *Vintage Bentleys are well versed at competing in today's motor sport events, this Three-litre attempts Honister Pass in Cumbria as part of the VSCC's Lakeland Trial in November 2004.*

Bottom Right: *When it was introduced in 1930 the Eight-litre was the grandest and most powerful of all Bentleys and rivalled Rolls-Royce's Phantom. Bentley Motors were, however, experiencing extreme financial difficulties owing to Britain's economic decline. This followed the Depression that started after the Wall Street Crash of 1929, so announcing such an expensive and prestige car only added to the firm's problems.*

Bentley's idea of building a sports car was conceived during World War I whilst working with Frederick Burgess, Humber's Chief Designer; the firm having been contracted to facilitate development of the BR1 and BR2. Bentley shared his ideas for a post-war sports car with Burgess, the formula being of around three-litres capacity and capable of providing travel over long distances with reliability and comfort. It was not until 1919 that Bentley's ideas came to fruition with the establishment of Bentley Motors: Burgess resigned from Humber to work for Bentley, and he was then joined by Harry Varley from Vauxhall. A small office in London's Conduit Street was rented, and it was there that the Bentley blueprint was created. Building the experimental cars was undertaken at New Street Mews, although the premises lacked the space for serious production. A search for alternative works resulted in the purchase of land at Oxgate Lane in Cricklewood, north London, on which a purpose-built factory was constructed.

In WO's words, his car, to be known as the Three-litre, should be "A good car, a fast car, the best in its class." It proved an excellent machine, beautifully designed and engineered to rival the most exclusive marques of the day. By comparison to some car-makers, Bentley Motors' workforce may have been small, but each person was handpicked by WO. He was a fair man with great integrity, nevertheless he instilled strict discipline within the factory, and any employee seen misusing tools or contravening regulations was likely to be instantly dismissed.

Top Left: *The financial collapse of Bentley Motors in 1931 summoned a contentious take-over of the firm by Rolls-Royce who ultimately formed a new company, Bentley Motors (1931) Ltd. W.O. Bentley was thus denied the chance of designing a new car, the 3^1/$_2$-litre, which was built at Derby, an example of which is pictured here in England's West Country.*
Oliver Suffield

Bottom Left: *Pictured near Malvern in 1936, the year the 4^1/$_4$-litre Derby Bentley was introduced, this Thrupp & Maberly-bodied car presents a fine image. The car, carrying registration number JB 1000, was (at the time) Bentley dealer Jack Barclay's demonstration vehicle.*

Bentley clientèle were wealthy and demanding, comprising the professional classes, aristocracy and royalty. As a company Bentley Motors was severely under capitalised from the very beginning. Resources in respect of tools and equipment were limited, and it was years before the firm had its own machine shop. Had it not been for the loyalty of the workforce, customers and shareholders who bolstered the firm's fragile finances, the company would have gone out of business within months of becoming established. WO maintained a high sporting profile with the Bentley Three-litre: it quickly acquired a reputation as being a sports car of the highest quality, an eminence that was confirmed in competition events, none of which was more demanding than the Le Mans 24 hour race.

Despite receiving good publicity, Bentley Motors failed to prosper financially. Although a brilliant engineer, WO was not a businessman, and the possibility of Bentley Motors going into liquidation in 1926 was very real. A solution aimed at keeping the firm financially afloat meant receiving pecuniary backing from millionaire racing driver and Bentley customer, Woolf Barnato, who became Bentley Motors' Chairman in place of W.O. Bentley. Nevertheless Bentley was retained as Racing Manager and Technical Director, positions he held until 1931 when the firm went into receivership following withdrawal of Barnato's investment owing to the downturn of Britain's economy.

Though WO had intended his cars to be open tourers, many customers preferred coachwork that was considerably heavier than had originally been envisaged, thereby compromising the Three-litre's performance. The demand for more commodious coachwork led to introduction of more powerful cars, a 6^1/$_2$-litre six-cylinder, a 4^1/$_2$-litre four-cylinder, and the ultimate 'WO' Bentley, the Eight-litre.

Top Right: *Wearing Park Ward drophead coachwork, this 4¹/4-litre car was delivered to its first owner Dennis Foden of ERF Trucks Ltd, on 1st June 1939. Despite W.O. Bentley's disappointment at not being asked to design the Derby cars, he nevertheless graciously acknowledged them to be very fine vehicles.*

Middle Right: *Derby Bentleys were fitted with coachwork by the most respected coachbuilders, this 3¹/2-litre car wearing a Barker fixed head coupé body. A Bentley publicity photograph, the car is seen near Shere in Surrey. The car's first owner was the Fourth Baron Fermoy.*

Bottom Right: *Pictured as recently as 2000, this October 1936 4¹/4-litre four-door saloon has coachwork by Park Ward. The larger engines had been adopted to improve performance and compensate for heavier and more commodious types of coachwork.*

With financial disaster looming, Bentley Directors decided to introduce a new model utilising a chassis similar to that of the Eight-litre but with a four-litre engine. WO would have nothing to do with the car, and only fifty were built before Bentley Motors collapsed. However, as the Marque had such prestige, it was unlikely to be allowed to disappear without trace. Napier, having ceased motor manufacturing in 1924 in preference to building aero engines, sought to acquire Bentley's assets as being their means of returning to car making. This was not to be, because the sale of Bentley to Napier was thwarted in the courts following a hostile bid made on behalf of Rolls-Royce before the offer could be ratified.

In acquiring Bentley Motors, Rolls-Royce closed the Cricklewood factory but inherited the services of W.O. Bentley. The relationship between the Directors of Rolls-Royce and WO was far from ideal, and though Bentley Motors (1931) Ltd was formed as a separate company, WO was not invited to take up a Directorship. All the more insensitive was the demeaning attitude towards WO, depriving him the opportunity of designing an all-new Bentley motorcar. Somewhat patronisingly, WO was invited to undertake testing of the new Bentley six-cylinder 3¹/2-litre, which was built by Rolls-Royce at Derby to become known as 'The Silent Sportscar'. As a result W.O. Bentley's tenure with Rolls-Royce was short-lived, and in the mid-1930s he took a position with Lagonda as Technical Director. There he designed the Lagonda LG45 and the outstanding V12, the latter winning its class at Le Mans in 1939.

Top Left: *By the middle 1930s, styling issues were being dominated by streamlining as depicted in this view of an unidentified Bentley pictured at Oberau in the Bavarian Alps. The image also confirms the Bentley's touring potential. W.O. Bentley had always insisted in subjecting his cars to long test journeys across Europe covering hundreds of miles at a time.*

Bottom Left. *This deftly streamlined car (chassis B29LE) has coachwork by Vanvooren of Paris and is similar to the 4¼-litre Bentley designed by Georges Paulin and built by Paris coachbuilder Pourtout for racing driver André Embiricos in the late 1930s. The car pictured was delivered to Major Eric Loder in November 1938.*

WO endorsed Derby Bentleys as being fine motorcars; like their Cricklewood predecessors they were furnished with bespoke bodies courtesy of the finest coachbuilders of the day, and were sought after by a discerning clientèle for whom only the best was sufficient. The 3½-litre was replaced in 1936 with the more powerful Bentley 4¼-litre, a car acclaimed for its fast acceleration and exceptional smoothness.

During the mid-1930s there was much debate amongst Rolls-Royce management towards rationalising production by utilising, where possible, common chassis and engine designs throughout the model range. The architects of the rationalisation policy were Ernest Hives, General Manager of Rolls-Royce, and Chief Engineer William Robotham, and the first Bentley model to appear in this respect was the Mk V in 1939. Of the 35 Mk V chassis that were laid down (50 were sanctioned), only eleven cars were built before production ceased at the onset of war. The remaining chassis were used for experimental purposes, or destroyed.

The Mk V was the true precursor to the post-war Bentley, and moreover formed the basis of the sporting derivative that emerged in 1952 as the Bentley Continental. Compared to the 3½ and 4¼ litre chassis, that of the Mark V with its diagonal bracing and deeply-formed side-channel-section members proved extremely durable. The design incorporated Packard-derived independent front suspension, a first for Bentley but already used on the Rolls-Royce Wraith.

Completely new was the 4257cc inlet over exhaust engine, which had Rolls-Royce influence in respect of its timing arrangement, tappets and main bearing design. A new gearbox was devised, with synchromesh on second, third and top ratios, rather than third and top as previously. Unchanged, though, was the traditional right hand gear selector.

An interesting development in preparing the Mk V was evaluation of a front torsion bar suspension, a feature seen on Citroën's radical Traction Avant of 1934. Rolls-Royce engineers ultimately abandoned the idea of torsion bars in favour of coil springs and triangulated wishbones.

The onset of hostilities in 1939 meant that the Derby factory was given over entirely to aero engine production. Remnants of the car division were relocated to a disused iron foundry at Belper in Derbyshire, and experimental cars, which would have heralded those models intended for introduction in the early-1940s, were dispatched to Canada for the duration of the war. The blueprints pertaining to future designs were taken to Ashby-de-la-Zouch in Leicestershire, where they were deposited in a bank vault for safe-keeping.

A number of prototype Bentleys were retained for use by senior Rolls-Royce personnel as well as being loaned to various dignitaries; one particular car, an experimental drophead coupé, being supplied to Air Marshal Sir Arthur (Bomber) Harris. Experimental cars were essential, for they gave engineers based at Belper some idea of modifications to be applied when car production resumed.

Above: *In 1939 Rolls-Royce underwent a programme of rationalisation in respect of Rolls-Royce and Bentley cars, the latter being represented by the MkV, of which only 11 were built before the onset of war.* Rolls-Royce

During 1943-4, Rolls-Royce reviewed its future aero engine and car manufacturing activities. Aero engines were to be located at Derby, leaving the Crewe works, built in 1938 under the Shadow Factory Scheme to make Merlin and Griffon aero engines, to take over car production. As aero engine work was being scaled down, the factory was converted to car-making.

The cars that had been sent to Canada for safe keeping returned to Britain in late August 1944, among them being a Mk V known as Comet but which received the nickname Scalded Cat owing to the vehicle's exciting performance. Rather than the six-cylinder engine fitted to production Mk Vs, a 5.3-litre cast iron in-line eight-cylinder engine was installed that was normally reserved for commercial and military applications. When a 6.5 litre straight-eight engine was fitted, acceleration was dramatically improved, reducing the 10-90mph time from 34.2 seconds to 28.1.

Top Left: *It is believed that of the 11 MkVs built, only nine survive. This H.J.Mulliner car was sold in July 1945 having been used on Rolls-Royce and Bentley business during wartime.*

Middle and Bottom Left: *New Bentley designs building on the 'rationalisation' theme were largely abandoned when war broke out in 1939. Some work did continue, however, the car illustrated having an eight-cylinder engine to provide 100mph top speed was shipped to Canada in 1940 for the duration of war! When it returned in 1944 it served as a company vehicle, and was loaned to several dignitaries, including HRH Prince Philip who was reluctant to part with it. The car's performance was such that it evoked the nickname 'Scalded Cat'.* Rolls-Royce

In addition to the Scalded Cat being used for experimental purposes, the car was also loaned to select V.I.P. customers. A valued client to test the vehicle was HRH Prince Philip, the Duke of Edinburgh, who appreciated its virtues to the extent that it was with the greatest reluctance that the car was eventually returned! Nevertheless, the Scalded Cat was responsible for Prince Philip persuading Rolls-Royce to build a state limousine for Princess Elizabeth, which was delivered to Her Royal Highness in 1950. A second Scalded Cat was commissioned during the immediate post-war years but with styling predicting the definitive Crewe Bentley. This was a handsome vehicle with a saloon body and twin side-mounted spare wheels, its straight-eight engine concealed beneath a long bonnet.

While the Crewe factory was being converted for post-war car production, the development of Bentley's 'new' range of cars was undertaken at Belper. Under the direction of William Robotham, Chief Stylist Ivan Evernden prepared drawings for a design that was considered modern but sufficiently conservative to be acceptable to Bentley cliéntele. Once approved by management, a full-size wooden model was constructed for endorsement by the Board of Directors.

Approval having been given to Evernden's designs, the next stage was to fully implement the rationalisation scheme which had been introduced before the war. Instead of all Bentleys being furnished with coachbuilt bodies to customer specification (it was appreciated that a select few of Bentley's clientèle would insist on having 'coachbuilt' cars), Robotham envisaged the mainstay of production having a standardised body design to be built by Pressed Steel of Cowley.

The Rover Company had already commissioned Pressed Steel to supply it with high quality bodywork, and was reaping substantial savings despite a high initial tooling cost. It was with some confidence, therefore, that Robotham set out to convince Directors this was the route Rolls-Royce should take for Bentley.

Robotham had been correct in his perception that the bespoke coachwork industry faced decline, and that the number of specialist coachbuilders having the capacity to produce vehicle bodies in volume had seriously diminished. To exacerbate the situation, many of the coachbuilders that had been in business in 1939 failed to resume operations after the war.

Discussions between Robotham and Pressed Steel began in January 1944: the firm was eager to win the prestigious Bentley (and later Rolls-Royce) order, and negotiated a production run of 2,000 bodyshells annually, the least number that was economically viable though somewhat greater than Robotham had envisaged. Pressed Steel's tooling costs were dramatically higher than Robotham had anticipated, but at £0.5m (£9m today) the figure of producing a standard-bodied car (these became known as standard steel saloons) represented approximately half of that of building a specialist coachbuilt vehicle.

Above: *Successful trials with the 'Scalded Cat' prompted Rolls-Royce to experiment with a second car having an eight-cylinder engine. Designed when plans were being made for post-war Bentleys to have pressed steel bodyshells built by the Pressed Steel Company of Oxford, Scalded Cat Two wears coachwork similar in style to that ultimately adopted.* Rolls-Royce

On the signing of the contract, Pressed Steel undertook the tooling for Evernden's design of coachwork, which was codenamed Ascot, and estimated that deliveries would commence during 1946. Prototypes of Bentley's first post-war car, the Mk Vl, were running at Crewe in February and March 1946 in advance of production commencing later that year.

Working with Ivan Evernden was John Blatchley, a brilliant stylist who, before working for Rolls-Royce, had been Head Stylist with the noted coachbuilder Gurney Nutting. This firm had been responsible for the bodies of several Bentleys including the Speed Six, which featured in the famous race between Woolf Baranto and the Blue Train. Joining Rolls-Royce's car division in 1946, Blatchley was about to set his seal on Bentley design for decades to come.

LUXURY MOTORING IN THE AGE OF AUSTERITY

The Bentley MkVI was officially announced in May 1946. Greeting it with wide enthusiasm, the motoring press was particularly complimentary about the car's luxurious appointment and superlative comfort; features missing from most cars during the period of post-war austerity. Also worthy of significant comment was the vehicle's ample reserve of power, its pleasing appearance, supple ride with absence of noise and vibration, lightness of steering, powerful braking and smooth transmission. All in all, the new Bentley represented British engineering at its best.

The car unveiled to the media was the first of the prototypes to carry Pressed Steel's standard steel coachwork, and its styling, fit and finish was much admired by everyone who examined it. This was, however, the second Crewe built MkVI, the first having been an H.J. Mulliner-bodied car with long front wings that swept downwards almost to sill height before blending into the rear wings.

Above: *The first post-war Bentley was the MkVI, the majority of examples having Pressed Steel's 'standard steel' coachwork. An early production vehicle, the extent of its luxury appointment can be gauged from the sumptuous looking hide seats.* Rolls-Royce

To no-one's surprise, this coachbuilt car tipped the scales some 225lbs heavier than the standard steel car, resulting in a top speed of 5mph lower than the standard car's 88mph. The standard steel car was used as a mule to produce trim patterns for production cars, the first of which left the assembly line on 21st September 1946.

The fact that Bentley cars spearheaded Crewe production, rather than the Rolls-Royce models bearing the Spirit of Ecstasy was a determined marketing ploy by the company who anticipated the Bentley monogram as being outwardly less ostentatious than the famous Grecian Temple radiator. In post-war austerity, where the nation had to 'tighten its belts', Rolls-Royces and ration books were hardly synonymous.

Top Right: *This scene of activity was pictured not at Crewe but at Park Ward coachbuilders in London, where final preparations are being made to this delectable MkVI drophead coupé. Note the car's doors are front-hinged rather than being rear-hinged as on standard steel saloons.* Rolls-Royce

Middle Right: *The Crewe factory in the early days of car manufacturing. The MkVI in the foreground has been through road test, hence fitment of registration plate and brightwork fittings.* Rolls-Royce

Bottom Right: *Every Bentley leaving Crewe was given a thorough examination and polish, each of the personnel making sure the vehicle in their charge received the best attention.* Rolls-Royce

The decision to buy bought-out bodies from Pressed Steel was a brave move by William Robotham, but one that proved to be highly beneficial for Bentley and its parent company Rolls-Royce. Whilst Robotham was convinced that overall quality would not be compromised by adopting proprietary coachwork, not all loyal Bentley customers shared his optimism. However, fears that a car with standard steel coachwork could not achieve an excellence similar to that produced by a specialist coachbuilder were unfounded; indeed the finish and appointment of standard steel cars proved to be every bit as good as their coachbuilt counterparts.

Supplies of steel in the aftermath of war were inconsistent and subject to varying differences in quality, a malaise that generally affected all Britain's car makers. The gauge of metal used by Pressed Steel for construction of Bentley bodyshells was the heaviest then available, and is one of two prime factors resulting in the healthy survival rate of the MkVI. The other is the very high standard by which the cars were constructed.

Bentley's post-war Crewe operation was vastly different to that employed at Derby pre-war. Whereas Bentley previously produced chassis for clothing by specialist coachbuilders, mass-production of bodyshells and building complete cars meant that output and financial turnover could be increased.

Producing complete cars gave rise to adopting totally different practices not previously undertaken by Bentley or Rolls-Royce and, as such, was the greatest step forward in the 40 year history of the latter company. Pressed Steel bodyshells were delivered to the Crewe factory where, on arrival, they were meticulously examined; any imperfections being rectified in-house.

Top Left: *H.J. Mulliner showed this MkVI at the 1949 Earls Court Motor Show. Note the streamlined coachwork which was constructed using lightweight 'Reynolds Metal' extrusions for the body framing. It therefore predated the Continental, which at the time was under development.* Rolls-Royce

Middle Left: *Bentleys galore! A collection of coachbuilt and standard steel cars at an enthusiasts' rally. The car nearest the camera is a Continental.*

Bottom Left: *This 1951 MkVI standard steel saloon was pictured at Kirkby Stephen in Cumbria in 2004, its splendid condition reflect the car's original build quality and its use of a high grade of steel.*

Preparing bodyshells prior to production was a time-consuming and labour-intensive process, involving provision for left or right hand drive (according to market destination), rust inhibiting, painting, and mating to the chassis. Factory personnel applied themselves to crafts previously undertaken by coachbuilders, and also embraced new skills such as upholstery trimming and decorative woodwork.

The quality and attention to detail applied to Crewe cars at this time formed the basis for production over ensuing decades. The procedures adopted were aimed at producing the finest cars that the British motor industry could offer, and also giving levels of craftsmanship that would have been difficult to match elsewhere. Such quality, nevertheless, reflected on the MkVI's price, which on introduction was a little over £4,038 inclusive of purchase tax at the rate of 66% on cars costing more than £1,000.

Whilst it was anticipated that the majority of Bentley customers would select standard steel cars, those who demanded bespoke coachwork were not neglected. It took longer to produce a coachbuilt car than it did a standard saloon, and therefore prices were considerably higher. The interior appointment of a coachbuilt Bentley car was usually more elaborate than the already lavishly equipped standard saloon, and customers could specify materials and styling features to individual taste.

Park Ward, incorporated within Rolls-Royce since 1939, offered coachbuilt cars from £4,964, while H.J. Mulliner and James Young priced their vehicles from £5,314 and £5,455 respectively. To put Bentley MkVI prices into perspective, an Alvis Fourteen saloon cost £1,276, and a Jaguar 3.5 litre £1199, whilst a Morris Eight commanded under £400.

In describing the standard steel MkVI's body arrangement, in 1946 *The Autocar* claimed it as having a "clean-cut style of its own, graceful and characteristic." This in effect meant the car had semi razor-edge styling, which gave the shape a formal element. The standard steel saloon that went into production benefited from some deft styling modifications introduced by John Blatchley to that originally proposed by Ivan Evernden. Among them was clever provision of concealed door hinges which operated on slides, thus opening doors away from the body. Most obvious about the Mk VI was its stately radiator crowned with the Bentley mascot.

Early models had ten shutters on each side, as against nine on later types. Lucas headlamps (which were partially faired into the front wings) had the 'B' motif behind the glass, while a single fog lamp was fitted to the central apron. Some coachbuilt cars, incidentally, had twin fog or 'passing' lamps located on side aprons off-set from the headlamps. The long bonnet concealed the 4257cc six-cylinder in-line engine, a beautifully refined creation that was impressively quiet when running.

Having an overall length of half an inch under 16-feet, the MkVI was a car of generous proportions. The chassis had a channel section frame and was of riveted construction, added strength being derived from cruciform centre bracing. Supporting the front suspension components, comprising coil springs, rubber bushed wishbones and double-acting hydraulic dampers, together with anti-roll bar, was a box-section pan. At the rear there were controllable hydraulic dampers, and leaf springs protected from dirt and moisture by leather gaiters.

Top Right: *The styling of the MkVI is shown to good effect here. Evident is the size of the car, its lofty driving position commanding a feeling of supremacy whilst at the wheel.*

Middle Right: *A Bentley MkVI alongside a Citroën Traction Avant, both cars having the same year of manufacture. The Citroën was introduced in 1934, the Bentley in 1946. Both cars are revered for their engineering and handling characteristics. W.O. Bentley, incidentally, drove a Traction Avant on business and was impressed by its performance.*

Bottom Right: *Not all Bentley owners were too keen about the MkVI's limited boot space, nor the drop-down boot lid, which was known to allow the ingress of water through the seal to collect in the spare wheel compartment; access to this was gained via the pull-out and drop-down panel incorporating the car's registration plate.*

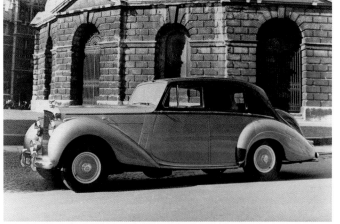

Top Left: *For the 1953 model year John Blatchley gave the MkVI a new lease of life by introducing some restyling to produce a larger boot and longer rear wings. The chassis length was also extended and suspension upgraded. The new lift-up boot lid was made from Birmabright alloy.* Rolls-Royce

Middle Left: *In restyled form the MkVI became the R-type. Note the extended rear wings and larger boot, the latter providing for much improved luggage capacity. The new boot lid design prevented ingress of water to the car's interior.*

Bottom Left: *This coachbuilt R-type car was for sale at the autumn Beaulieu Autojumble in September 2004, where it was the cause of much interest.*

Steering was by Marles-type cam and roller to allow three-and-a-half turns of the steering wheel lock to lock. A centralised 'one-shot' chassis lubrication system provided oil to all steering and suspension systems bearings, this being driver activated at 100-mile intervals via a pedal located beneath the facia. Yet, few drivers operated the system correctly, which was one reason why this apparently good idea was ultimately abandoned.

Meanwhile the braking system was illustrative of Bentley's 'belt and braces' engineering: so as not to be totally reliant upon new technology, the front drums were hydraulically controlled via a servo on the gearbox. The rear drums were mechanically activated via the foot pedal and servo, and a pull-type handle beneath the facia to the driver's right on right-hand-drive cars operated the parking brake. The rear-mounted fuel tank had an 18-gallon capacity, and at the heart of the electrical system a 55 amp/hour 12 volt positive earth battery was originally specified.

Travelling in a Mark VI was a delightful experience in the climate of immediate post-war Britain, and it remains thus. So it was for the many customers to whom the car was exported, the principal market outside the United Kingdom being Australia. The charm of the Mark VI has not diminished over the intervening years, the vehicle still being a delight to drive. Sitting at the wheel of a Mk VI one is conscious of the lofty driving position with its commanding view over the long bonnet to the radiator which is capped by the 'Flying B' and flanked either side by substantial, yet elegant front wings.

All this was sufficiently reassuring, and greater confidence was derived from the large three-spoke steering wheel, which was surprisingly light and balanced considering the car's size and weight. The Bentley insignia on the instruments and brake and clutch pedals is a gentle reminder of the car's heritage.

The effort that went into achieving levels of luxury and sophistication seldom seen on other makes of vehicle was quickly appreciated. Supple but supportive, the individual front seats were furnished with finest Connolly hide, though a bench seat could be specified. The unobstructed floor with its Wilton carpeting added to the interior spaciousness and allowed the driver to slide across the car to exit adjacent to the pavement rather than stepping out amidst passing traffic. Indeed, the sole external door lock was on the front passenger door.

The facia exuded craftsmanship, the finely veneered walnut instrument board being conservative, arguably to the point of being old-fashioned. Centrally positioned, the instrumentation comprised three dials, that in the middle being the switchbox, a familiar Bentley feature. To the right of the switchbox with its ignition master switch, starter press and light switches, was the speedometer; to the left a dial contained an ammeter, water temperature, oil pressure and fuel gauges.

Vertically placed at the extremities of the instrument board, switches controlled the windscreen wipers (additional control knobs on the capping rail were to engage and park them), instrument lighting, cigar lighter, heating and demisting. A wind-back switch fitted to the centre of the capping rail operated the semaphore direction indicators located in the centre door pillars. Many more modest cars lacked such luxuries as an efficient heater: that on the MkVI was located beneath the front passenger cushion and was fed with hot water from the engine cooling system. Activated by a switch on the instrument board, warm air was directed through vents between the front seats

Below: *An R-type standard steel saloon pictured in company with a Ford Consul and Standard 8. Shortly before introduction of the R-type, the MkVI was given an increase in power with a larger engine, the capacity up-rated from 4257cc to 4566cc.* Rolls-Royce

Top Left: *Whilst the R-type standard steel saloons were impeccably styled and constructed, there nevertheless existed a market for cars having bespoke coachwork with individual interiors, such as this James Young two-door saloon.*

Middle Left: *The facia and instrumentation as seen on a typical MkVI and R-type saloon. The fact that there was a single slide control on the steering wheel hub indicates this as being an R-type Bentley.*

Bottom Left: *This R-type drophead foursome coupé has coachwork by Park Ward, the folding head was power operated, whilst wide doors and sliding bucket-type front seats make the rear compartment easily accessible.*

Provision of a lockable glovebox, HMV radio, clock, automatic interior courtesy lights and a sliding steel sunroof were period luxuries. Beneath the facia in front of the passenger seat, a pull-out tray housed an array of small tools. Under the bonnet there was an inspection lamp, grease guns and an oil syringe. Completing the tool kit, tyre levers, hub spanner, Dunlop screw jack, wheelbrace and tyre pump were housed in the boot. Three sliding controls were fitted to the steering wheel hub: one was the fuel mixture, another the hand throttle with a cold start setting, the third adjusted the rear suspension damper settings to provide an optimum ride in respect of road conditions and vehicle weight.

Rear passengers were cosseted with deeply cushioned seats and a central armrest; ashtrays were located in side cushions. Incorporated within the rear pillars, illuminated 'companions', one having a cigar lighter, captured the grandeur of past generation grand tourers. As well as affording privacy, a rear window blind served to prevent headlamp glare from following vehicles, whilst carpeted footrests and folding picnic tables completed an imagery of total luxury.

By modern standards the MkVI's six cubic foot boot had meagre capacity, though additional luggage could be supported on the drop-down boot lid when required. The spare wheel was located in a separate compartment beneath the boot, access to which was gained from a removeable panel housing the car's number plate and tail and reversing lights. Sedate would be the best way to describe the MkVI's performance by today's standards, but at the time it was really quite outstanding. The soft ride allowed some roll on cornering; braking was adequate rather than exceptional in light of modern brake technology, but it required a determined effort in an emergency stop.

With petrol consumption no better than 17-18mpg, the MkVI was hardly economical, but such poor fuel efficiency meant very little to those people who could afford to buy a Bentley during the post-rationing period of the early-1950s.

A number of technical modifications were made to the car's specification during its career, the most significant being an increase in engine size in late-1951. By enlarging the bore from $3^{1}/_{2}$ to $3^{5}/_{8}$ inch, thus raising the swept volume to 4566cc, and power to give around 153bhp, the exact figure never officially revealed anything other than being 'adequate'.

The MkVI didn't escape criticism, and in 1951 John Blatchley re-styled the car to improve its boot capacity. He did this without dramatically altering the car's image, but nevertheless the modifications called for a new model designation, to be known as the MkVII. This modified Bentley was ultimately known as the R-type when introduced in September 1952, one reason being that Jaguar had by this time introduced its MkVII sports saloon. The MkVI chassis was lengthened by six inches and allowed for the rear suspension to be modified by widening and extending the springs. The fuel tank was located farther forwards, though its 18-gallon capacity remained. The revisions affect the car's styling rearwards from the B/C post by elongating the rear wings and expanding the boot to $10^{1}/_{2}$ cu ft capacity, the boot lid being top-hinged and formed from Birmabright aluminium alloy instead of steel.

Blatchley's revised styling was greeted with enthusiasm from customers who considered the R-type's profile to be improved aesthetically, and because water no longer seeped into the spare wheel or luggage compartments. There were other technical modifications: improved petrol quality obviated the need for a mixture control, and an automatic choke was fitted for cold starting, thereby eliminating two of the slide controls fitted to the steering wheel hub. Modified windscreen wipers meant it unnecessary to have manual engage and disengage control, and the small tool kit was relocated beneath the driver's seat.

Top Right: *This view shows the rear compartment of the MkVI and R-type; note the hide trim, veneered picnic tables and the rear quarter companion, which had a mirror, light and cigar lighter. Footrests and carpeting add to the luxury.*

Bottom Right: *It can be difficult to identify an R-type saloon, as shown here, from a MkVI, the cars' frontal aspect being virtually identical. A side profile would indicate the R-type's longer chassis length, extended rear wings and boot and, arguably, John Blatchley's more balanced styling.*

Top Left: *At first glance this R-type with Abbott coachwork could be confused with a Bentley Continental. Abbott-bodied cars are often wrongly described as Continentals, the latter commanding vastly higher values.*

Middle and Bottom Left: *During the early to mid-1950s work began developing an entirely new Bentley that was not introduced until 1965. Here is an experimental 'Burma' car on test in France at Le Mans, test driver John Gaskell is taking a welcome break. The lower picture depicts the definitive car, the Bentley T-series, on its media launch.* Rolls-Royce

With the R-type came the option of an automatic gearbox, which initially was specified for export orders. The first time an R-type was exhibited with such transmission was at the 1953 London Motor Show, and soon afterwards was standardised though manual transmission remained optional until production ceased in 1955. The automatic gearbox fitted to the R-type was built in America by General Motors, though it was modificd to meet Bentley requirement, and therefore unique. Manufacture of the 'Hydramatic' gearbox was allowed at Crewe under licence. Modifications meant that not only could it be manually actuated via the column-mounted, rather than the more usual floor selector, a second-speed hold device was designed to provide engine braking when descending steep hills. A safety override was fitted to the third speed, so that it automatically selected top gear should a car's speed exceed 60mph; an arrangement designed to minimise engine speed and wear.

During the latter stages of R-type production, work on its 1955 successor, the S-series, was being finalised. By then thoughts were already engaging on what was to follow the S-series, and in this respect Blatchley was adamant that a great stride lay ahead. Advancing technology meant Rolls-Royce and Bentley had to adopt new practices, and it was his aspiration to bring both marques right up to date with a car employing monocoque construction and being completely innovative. This giant step, however, remained more than a decade away.

Production of MkVI and R-type standard steel saloons accounted for 4,188 and 2,037 cars respectively. Of the specialist coachbuilt cars there were 1,012 MkVIs and 283 R-types, the most selected coachbuilder being H.J. Mulliner who completed 368 examples, 67 of them R-types. The other popular coachbuilders were James Young (209 MkVIs and 69 R-types), Park Ward (167/50), Freestone & Webb (103/29) and Hooper (61/41). A total of 7,520 MkVI and R-types were built, the coachbuilt examples today commanding high values.

JOHN BLATCHLEY'S MASTERPIECE

The S-series Bentley is regarded as being John Blatchley's masterpiece, and the design even inspired styling cues for the Bentley Arnage that was introduced in 1998.

The S-series had already been conceived on Blatchley's drawing board when the R-type was unveiled in 1952, having established the early design principles at least a year previously. Bentley's records indicate that the S-series was initially going to be known as the MkVIII, and at least two prototype vehicles were built with coachwork showing all the hallmarks of what eventually emerged as the definitive shape. The MkVIII programme was abandoned in favour of the MkIX, a concept code-named within the Crewe factory as 'Siam'. Since the MkVII had evolved as the R-type, so its successor adopted the 'S' designation.

Above: *The Bentley S-series was introduced in the autumn of 1955. From whatever angle the car is viewed, it looks impressive and substantially larger than its predecessor. Like the MkVI and R-type, the S-series was designed to carry Pressed Steel's standard steel coachwork.*

The success of the MkVI and R-type models was due in no small degree to the high standard of the bodywork by Pressed Steel Ltd. and therefore, from the outset of the Siam programme, it was agreed that this would be the sensible route for the S-series to follow. For those customers who sought the ultimate in coachwork, the S-series chassis would also be made available to specialist builders, though fewer firms remained in business by the mid-1950s than had previously been the case. By 1955 many old established coachbuilders had gone out of business, while the few survivors remained trading as best they could.

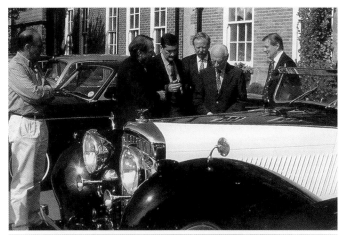

Top Left: *John Blatchley (centre) on his 90th birthday reflecting on the styling he designed. The car in question is a Derby-built car with Gurney Nutting coachwork.* R-REC

Middle Left: *An experimental S-series saloon seen at Crewe. Note the large rear fins fitted to test the car's side-wind stability. Testing cars at Crewe was an on-going exercise aimed at improving technology.* Rolls-Royce

Bottom Left: *The S-series Bentley in quarter-scale form in the Crewe styling studio. A fibre glass shell, it reflects styling trends that led to John Blatchley's design masterpiece.* Rolls-Royce

The S-series signalled the beginning of an entirely new era for Bentley: based upon post-war conceptions the genus had little or nothing to do with models that had been conceived pre-war, other than maintaining the tradition of highest quality engineering. In designing the S-series, Blatchley worked in close liaison with Chief Engineer Harry Grylls and Managing Director Dr. F. Llewellyn-Smith. In those days there were no styling committees, only those people having experience of precisely what was needed; history rarely proving them wrong.

For Bentley, the S-series was a landmark car in respect of styling technique, for the final shape was the first to be created from a quarter-scale clay model. This technique revolutionised the firm's styling practice, and for the first time allowed for a super-accurate model to be produced, and when approved for production could be measured and drawn to scale. Before styling clay became universal, Blatchley had introduced modelling in Plasticine. Although it was cheap and convenient, working with this material had disadvantages since it was oil-based and had a tendency to soften in warm temperatures. Clay was a different matter, and once warmed in a small electric oven to become malleable, it was made to cover a plywood box, known as the 'coffin', that was slightly smaller than the proposed project giving a final clay thickness of something like a quarter to half an inch. Once the clay had hardened, stylists shaped it using sharpened household paint scrapers to the desired profiles, after which detailed measuring process began.

The design Blatchley proposed for the S-series Bentley was approved by the Directors in 1951, and the first experimental vehicle was delivered to Crewe from Rolls-Royce's in-house coachbuilder Park Ward in September the following year. The car's coachwork had been handbuilt at Willesden to the same design as Pressed Steel's bodyshell which had been approved.

Above: *This early R-type saloon has coachwork by the firm of H.J. Mulliner and was delivered to its first owner, Strand Riley & Co. Ltd. in January 1953. Note how the coachwork styling differs to that of the standard saloon, particularly in respect of the front wings that sweep dramatically to meet the rear wheel arches. Note also the razor-edge profile, which is more severe than on Pressed Steel's coachwork.*

Right: *A collection of MkVI and R-type saloons. Customers often chose dual colour schemes for their cars at a time when most vehicles were delivered in a single colour, more often than not, black. R-type saloons differ from MkVIs in respect of an extended chassis, longer rear wings and a more capacious boot, which instead of having a drop-down opening is top-hinged. The engine is also of larger capacity to provide improved performance.*

Above: *Bentley MkVI and R-types enjoy a distinct presence, as seen by this H.J. Mulliner saloon pictured at Drumlanrig Castle near Dumfries. Familiar features of these cars were the stately radiator shell and sculptured front wings with 'semi faired-in' headlamps and wing-top sidelamps. Beneath the bonnet lay a finely engineered six-cylinder engine which could quietly and effortlessly propel the car to nearly 90mph.*

Left: *By contrast to the MkVI the R-type Bentley Continental had a reduced frontal area, faired-in headlamps and beautifully shaped wings. Exquisite was the car's streamlining, which helped it achieve in excess of 120mph whilst at the prototype stage. The car's ancestry can be traced to pre-war days, when Bentley designers were looking to create a sports saloon reminiscent of the grand tourers, as fashioned by W.O. Bentley. A total of 208 R-type Continentals were built, thus assuring exclusivity.*

Above: *Delivered in February 1955 to its American owner, this R-type Continental was among the last to be built prior to introduction of the S1 Continental. The majority of R-type Continentals, including this car, were constructed with H.J. Mulliner coachwork, the coachbuilder's expertise being recognised for its weight-saving technology. Today Bentley R-type Continentals have achieved iconic status and despite commanding high values are sought after by marque enthusiasts.*

Right: *Part of the pleasure of driving a MkVI or R-type is looking across the long bonnet to the Bentley mascot adorning the radiator cap. In their day these cars were amongst the most expensive to rival the like of Daimler and Armstrong Siddeley. Few cars equalled their comfort and interior appointment, the emphasis being on the quality of materials used.*

Above: *John Blatchley's masterly styling is clearly apparent in this photograph of the Bentley S-series, which was introduced in 1955. Four years later the six-in-line engine was replaced by a V8, which remains, in modified form, to this day. Designed by Jack Phillips, the V8 engine has since undergone many adaptations in accordance with advancing technology and emissions regulations; in its latest incarnation the engine develops 457 horsepower.*

Left: *The frontal styling of this Bentley depicts it as being an S1/S2 Continental. Differences between this and the previous model include repositioned sidelamps and higher, more prominent, rear wings. S1 Continentals were powered by the same 4887cc six-in-line engine fitted to saloon models, the S2 having use of the 6230cc V8. Bentley Continentals of the period were amongst the fastest and most expensive cars in the world.*

Top Right: *Emerging from a driveway leading to a stately home, this S-series Bentley evokes the image many people associate with such a car. When introduced, these Bentleys were available to a select clientèle, such was the cars' price of £4,669 inclusive of purchase tax.* Rolls-Royce

Middle Right: *Despite the high price of the S-series Bentley, the car found many customers abroad, this one being loaded aboard the SS* Brandager *en-route to America.* Rolls-Royce

Bottom Right: *The elegance of this S-series Bentley matches its surroundings, in this instance the headquarters of the Rolls-Royce Enthusiasts' Club, which caters for owners of Crewe-built Bentleys as well as all Rolls-Royces.*

By mid-1953 the experimental vehicle had been fitted with power-steering and air-conditioning, features that would later become standard. Another five experimental cars were built before the S-series went into production prior to its autumn 1955 launch. Within their development period many problems were addressed, the main concerns being road noise, body shake and a booming from the bodyshell. Problems with experimental cars were not uncommon, and only through continual trial and error could faults be overcome. The S-series' design incorporated few features from earlier models, but those that did make the transition were the rear axle and Hydramatic gearbox.

The chassis side members were of welded box-section, replacing the open channel section of the R-type, and whilst marginally heavier nevertheless had much greater rigidity. Rear semi-elliptic springs, still with leather gaiters (but upgraded from seven to nine leaves), were relocated outside the chassis frame side members, and at the same time were grease packed instead of being fed with oil from the central lubrication system, itself restricted to the front suspension.

In modifying the braking system Girling 'Autostatic' brakes gave the car impressive stopping power with surprisingly little effort on the brake pedal. Additionally, hydraulic actuation was extended to the rear drums, although in belt and braces fashion these retained their mechanical operation. Revisions to the front suspension allowed for changes to the car's steering geometry, now altered to require 4$\frac{1}{2}$ turns lock to lock instead of the previous 3$\frac{3}{4}$. This 'sneeze factor' was introduced by Harry Grylls who felt that an involuntary sharp movement of the steering wheel could lead to loss of control of the car if the steering was too highly geared. This was, however, a feature disliked by many, and was soon changed on Harry's retirement.

Above: *The Bentley S-series with its Rolls-Royce sibling, the Silver Cloud, outside the Crewe factory. The styling of the radiator shell and bonnet profile adds to the car's sporting appeal, which suited marque enthusiasts mindful of Bentley's racing history.*
Rolls-Royce

Motoring correspondents could be forgiven for perceiving the S-series to be somewhat old fashioned by virtue of its separate chassis, at a time when manufacturers of other luxury cars were changing to unitary construction. However, this was a large step that was not to appear at Crewe for another ten years. So for the S-series a contract was placed with John Thompson Motor Pressings of Wolverhampton, to supply conventional chassis frames for both standard-bodied cars as well as those with the optional four-inch longer wheelbase. In the case of the latter, cars would be either specialist coachbuilt or have specially lengthened standard steel bodyshells.

When the S-series was unveiled at the London Motor Show, it was not its specifications that impressed the media but the elegant coachwork. So appealing was John Blatchley's design that the few remaining bespoke coachbuilders found it difficult to improve the car's aesthetics, or for that matter to devise interiors that were more finely appointed. Appearing considerably larger than its predecessors, the wheelbase was increased by three inches compared to the MkVl and R-type, and overall the car was 12 inches longer than the former, all of which provided for generous interior spaciousness.

The bodyshell was of 20-gauge steel, the doors being formed from 18-gauge Birmabright aluminium alloy and bonnet and boot lid from similar 16-gauge material. The lower body was zinc-plated to protect against corrosion. The front wings were massive steel pressings incorporating sidelamps on the top edge, whilst the swage line swept gracefully down to the rear wings to afford a beautifully balanced profile.

The car's lavish interior featured a front bench seat with individually adjustable backrests, trimmed with Connolly hide. Veneered picnic trays built into the back of the front seats are standard features, as were companions built into the rear quarters. The facia with its walnut veneer facing was truly impressive: the instrument board was centrally located, the speedometer to the right of the switchbox, and to the left a dial incorporating a clock, ammeter, fuel, oil pressure and water temperature gauges.

Top Right: *Bentleys command much interest at vehicle sales, but peering into the bonnet of such a car might be proved somewhat intimidating, but compared to modern cars they are relatively simple.*

Middle Right: *S-series cars afforded the utmost comfort, and all this makes driving a Bentley a pleasurable experience.*

Bottom Right: *When the S2 Continental was introduced, the V8 engine was immediately acclaimed. Designed by Jack Phillips, the engine mirrored American trends and, contrary to popular belief, was not American built.*

Automatic transmission was standard on the S-series, though a few cars were supplied with manual transmission to customer specification. The ride control was retained, enabling drivers to adjust rear suspension damper settings, the operating switch was however relocated from the steering wheel hub to the left hand side of the steering column. In Bentley fashion the parking brake remained under the facia, outboard to the driver. Under the bonnet was a derivative of the R-type's six-cylinder engine (enlarged to 4887cc, 178bhp), which provided effortless motoring with a top speed of a little over 100mph and an average fuel consumption of 14.5mpg. A year after the car's introduction power assisted steering became optional, as did air conditioning (an American import), on export vehicles only. In 1958 Bentley offered its own unit for home and export orders, the American unit being retained for cars destined to the USA.

When the optional long wheelbase S-series was introduced in September 1957, standard saloon bodies were sent from Pressed Steel to Rolls-Royce's Park Ward coachbuilding facility for conversion. The process entailed cutting the bodyshell in two behind the B/C post and inserting four-inch sections to roof and floor, and fitting longer rear doors. With this body conversion complete, the shells were sent to Crewe for finishing.

In September 1959 a more powerful version of S-series Bentley was introduced, the important difference being an all-new 6¼-litre engine. The V8 was shorter and wider than the straight-six engine; it was also some 30lbs lighter. Yet, in order for it to fit within the engine compartment some shoehorning was necessary and this resulted in repositioning the steering box from within the chassis frame to outside it. The engine bay was occupied by a plethora of pipes, belts and hoses, and to illustrate the car's complexity it proved impossible to change the spark plugs from above the engine, a task involving removal of panels in the inner wheel arches.

Top Left: *This view of Crewe's styling studio depicts differing headlamp and frontal arrangements being tried on a scale model S-series. The modelling table allows models to be built to a precise scale. In the background can be seen styling engineer Martin Bourne.* Rolls-Royce

Middle Left: *The different styling arrangements seen in the previous photograph were part of the design process leading to introduction of the Bentley S3 for the 1963 model year. The four-headlamp system was popular at the time.*

Bottom Left: *The S3 successfully combined John Blatchley's classic styling with a measure of modernity to produce a most handsome Bentley for the 1960s. The car's modified frontal aspect employs a slightly lower radiator shell height, and sloping bonnet along with paired headlamps (a popular feature then emerging on American cars).*

Styling differences between the S2 and its predecessor were mainly confined to the car's interior appointment. The facia had a separate clock along with additional switchgear. To illustrate the S2's increased performance, the speedometer was recalibrated to indicate 120mph maximum speed.

A long wheelbase S2 was announced in September 1959 with deliveries commencing in January the following year. Bodyshell conversions were again entrusted to Park Ward. Other than minor mechanical differences the cars were similar to their predecessors though some interior modifications were made during the production span. A striking new appearance was given to the S-series in October 1962 in readiness for the 1963 model year.

The facelift mainly concerned the car's frontal appearance which had a twin headlamp arrangement with combined direction indicators and sidelamps incorporated into the leading edges of the front wings. Not immediately noticeable was the radiator shell, which was reduced in height by 1.25 inches, and demanded a bonnet of increased slope. The four-headlamp system was adopted because of its popularity in America and elsewhere, and it was also intended to give the car's overall design, considered by some to be 'slightly dated', a new lease of life. Under the bonnet, the V8 remained undisturbed although component modifications included larger throat carburettors, and an increase in compression ratio from 8:1 to 9:1.

A third derivative S-series Bentley was contemplated, the proposed S4 having hydropneumatic self-levelling to harmonise the otherwise conventional suspension.

John Hollings, at that time Rolls-Royce and Bentley's Chief Engineer, who had taken over from Harry Grylls on his retirement, decided that automatic self-levelling in conjunction with the S-series' springing provided the ultimate in ride quality. Despite efforts by Rolls-Royce to develop its own system, it was unable to match or better that used by Citroën, and therefore it was intended to use the French design under licence. Ultimately the S4 project was overtaken by development of a new and radical model which emerged as the Bentley T in the autumn of 1965

A total of 6221 standard steel S-series cars were built between 1955 and the autumn of 1965. Of these 3,072 were S1s while S2s and S3s accounted for 1,863 and 1,286 vehicles respectively. Of the long wheelbase cars there were 35 S1s, 57 S2s and 32 S3s. Relatively few coachbuilt examples were supplied: 157 were S1s, only 12 of which were delivered on the long wheelbase chassis; there were 20 S2s, five being long wheelbase cars, and of the eight S3s specified, only one was built on a standard chassis.

The elegance and quality of the S-series cars makes them sought after by discerning enthusiasts, John Blatchley's masterly design having an appeal which is in the truly classic idiom. Most collectors agree that it is the S2 which is the most desirable of the three models by virtue of its styling originality combined with the effortless energy of Jack Phillips' enduring V8.

Top Right: *The Bentley S-series shared coachwork design with its Rolls-Royce sibling, in this instance the Silver Cloud III. Bentleys and Rolls-Royces were built on the same production line at Crewe, differences between the standard steel saloons amounting mainly to radiator design and bonnet shape as well as badging. Inside the car a sense of luxury exuded with the finest veneers adorning the beautifully proportioned facia.*

Middle Right: *It was proposed to introduce a Bentley S4 with suspension incorporating hydraulic self-levelling but the model failed to materialise owing to ongoing and complex design technology involved with development of the T-series cars which were announced in the autumn of 1965.*

Bottom Right: *Many enthusiasts regard the S2 with its finely engineered V8 engine to be the ultimate Bentley S-series car. Relatively few coachbuilt cars were specified, such was the standard steel saloon's exquisite styling and build quality.*

CONTINENTAL MOTORING

When the Bentley Continental was introduced in 1952, it was not only the most luxurious sports saloon in the British motoring catalogue, it was, according to *The Autocar*, the fastest and most expensive production car in the world. At a time when many Britons were experiencing the effects of post-war austerity, there was however, a growing mood of optimism stemming from the Festival of Britain held the previous year.

Though initially intended for the important export market, the Bentley Continental found willing customers at home and abroad, the mere presence of this superbly designed car with its exotic styling and scintillating aerodynamic fastback helping to bolster national pride. In the same way that Britain's De Havilland Comet was at the cutting edge of aviation technology, so the Bentley Continental offered motoring in the grandest style.

Unlike the MkVl which was available with standard steel coachwork, this new high speed Bentley to be built in limited numbers was only ever specified as having bespoke coachwork.

Above: *The Bentley Continental's origins can be traced to the late 1930s when Rolls-Royce was looking to introduce a high performance Bentley sports saloon. This is the Corniche, an experimental Bentley with coachwork styled by Georges Paulin and constructed by Vanvooren of Paris. Whilst being tested in France the car was involved in an accident and suffered extensive damage.* Rolls-Royce

The Continental's origins lay deep within Bentley's pre-war programme, having been intended as an offshoot of MkV development. However appealing the Derby Bentleys were, there nevertheless lacked a model that had the sporting essence of WO's eight-litre car. The firm proposed to satisfy demand for such a vehicle, and in the late 1930s, design work began on a high-performance sports saloon intended for launch at the 1940 London Motor Show. Experiments commenced in 1938 with a lightweight car to be known as Corniche, the name synonymous with Mediterranean affluence. Instead of being designed in-house, the car was styled by the reputable French stylist Georges Paulin and constructed by Vanvooren of Paris.

Top Right: *A car that helped inspire the Continental's design was this machine for racing driver André Embiricos.*

Middle Right: *An experimental Corniche chassis pictured at the Derby factory.* Rolls-Royce

Bottom Right: *The Embiricos Bentley on test. The car, like the Corniche, was designed by Georges Paulin, and the streamlined coachwork built by Pourtout of Paris.*

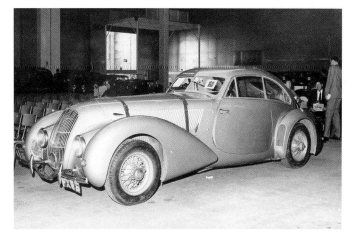

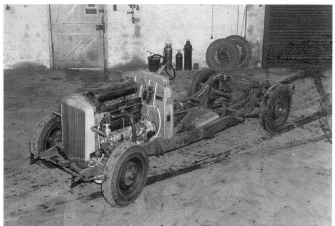

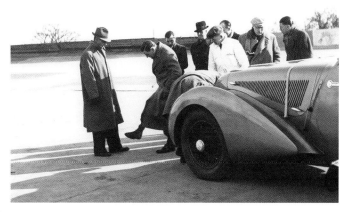

In order to keep the weight to a minimum, the chassis was formed from thinner gauge steel than normal and magnesium alloy castings were employed wherever possible. The Corniche did not look like a Bentley, and it was only the familiar insignia that identified it as such. Missing was the traditional radiator, in its place an art-deco grille, and there lacked the traditional expression of formality about the coachwork which adopted the latest ideas about streamlining. A cowled front end with integral headlamps and a uniquely shaped body with a narrow split vee windscreen gave the car its unique sporting presence.

Despite it demonstrating some stability problems at high speeds, the experimental car performed well during testing. Trials, however, were abruptly halted when the vehicle was involved in an accident in France, the damage being extensive. The car was taken to Rolls-Royce's depot at Châteauroux where chassis and body were separated, the former being returned to Derby for repair. The body was repaired in France but was then ironically destroyed during an air raid on Dieppe harbour. Georges Paulin was responsible for another distinctively streamlined 4$\frac{1}{4}$-litre Derby Bentley, a car specifically designed for Greek racing driver André Embiricos, which can also lay claim to influencing the Bentley Continental.

Forerunner to the Vanvooren bodied-Bentley, the Pourtout car was taken to mainland Europe in February 1939 where it was tested by John Dugdale of *The Autocar*. The results were dramatic, showing the car to cruise easily at 100mph and returning 20mpg at 80mph. In July the same year, in the hands of the one time land speed record holder Captain George Eyston, the car completed 114 miles in one hour at Brooklands. In 1949 it was entered in the Le Mans 24-hour race by its then owner, H.S.F. Hay to finish sixth, and happily this most significant Bentley survives to this day. With the onset of hostilities and subsequent destruction of the Vanvooren bodied car, the Corniche project was abandoned. The idea of developing a fast continental tourer in the genre of the pre-war grandes routières was not entirely forgotten though.

Above: *Standing alongside the prototype Cornice II, forever known as Olga owing to its registration number, is Ivan Evernden, at the time Rolls-Royce's Chief Designer who was responsible for the car's elegance.* Rolls-Royce

By 1950 the Corniche project, known as Corniche ll, had been resurrected. The idea for a car capable of 125mph came about through efforts to produce a more sporting version of the MkVl saloon, a project that did not, however, receive the total enthusiasm of Rolls-Royce Directors. Thankfully, within the management there were those who had sufficient wisdom to agree to developing such a car, the styling of which was entrusted to Ivan Evernden and John Blatchley, who at that time were working not from Crewe but Clan Foundry.

The earliest existing drawings of the proposed car were created between late 1949 and early 1950 to express John Blatchley's ideas for a two-door sports saloon with fastback styling, thus capturing the essence of the Pourtout bodied Embiricos Bentley. A year later the emergence of 1/16th scale drawings depict a revision of ideas which won approval for the car to enter limited production.

To achieve the performance demanded, it was essential that weight be kept to an absolute minimum, and this led to H.J. Mulliner & Co being awarded the contract to build a prototype car. H.J.Mulliner & Co. was noted for its particular expertise in the use of lightweight materials, and the firm's technical director, Stanley Watts, was assigned to preparing full size drawings prior to the car going into production.

Top Right: *The interior of the production Continental. Everything was done to achieve weight savings, sacrificing even the fitment of a radio.*

Middle Right: *Testing the Continental prototype summoned speeds in excess of 116mph, but when the car was taken to the Montlhéry circuit in France 120mph was achieved. The car's top speed during testing was a shade under 125mph. The Continental's aerodynamic shape was devised by Ivan Evernden and John Blatchley.*
Rolls-Royce

Bottom Right: *The Continental's rear compartment accommodated two passengers, though space was more limited than the picture suggests.*

Working with Stanley Watts was George Moseley, who was given the task of designing the car's all-metal lightweight structure. A quarter-scale model of Corniche II was then wind tunnel tested at Rolls-Royce's Flight Test establishment at Hucknall in Nottinghamshire. This testing was done to assess the quarter-scale model's aerodynamic drag and crosswind stability. Not only did the car's streamlining and fastback shape reduce drag, the destabilising effect of side winds was diminished because of the finned rear wings. The original drawings of the car depict a much lower rear wing line than that eventually devised, stability testing having proven that a raised wing line was more successful than the deeply curved shape that dropped gently down to meet the tail at bumper height.

Mulliner's construction technique utilised aluminium panels fitted to a built-up frame of extruded light alloy. Further weight savings included employing tubular construction for seat frames and using light alloy for window frames and bumpers. The design of the car called for a maximum weight of 34-cwt, which was dictated mainly by tyre technology of the time. The car finally weighed in at 33½-cwt, thus just within the limit. The prototype was finished in metallic grey and was sent to Crewe in August 1951. By this time the Continental appellation had replaced Corniche II.

Though the first series of cars are retrospectively known as R-type Continentals (the car was only marketed as the Bentley Continental), the prototype and earliest production vehicles were actually built on the MkVI chassis using the 4566cc engine rather than the 'small-bore' 4257cc unit. It was only after the release of the R-type Bentley that the modified chassis was used for the Continental.

Above: *The Bentley Continental was introduced in 1952, the same year that the De Havilland Comet made its debut on BOAC's Johannesburg service. At launch, the Continental was the fastest and most expensive production car in the British catalogue.* Rolls-Royce

Before its media launch, the Continental underwent extensive testing in Britain and France. Crewe's test drivers were only too keen to get their hands on Olga, and within a couple of days of the car's arrival at the factory, they were pushing it beyond 100mph. One of the test drivers' favourite stretches of road was King Street, a long and very straight but quite narrow road linking Northwich and Middlewich. It was on here one day that the car first achieved 116mph.

Within a week of being delivered to Crewe the prototype was dispatched to France where tests were conducted at Montlhéry, the scene of many record breaking attempts during the 1930s. It was there that 116mph was again recorded, and in an attempt to push the speed still higher, minor damage to the engine was sustained when it overheated owing to water pump failure.

Subsequent testing resulted in 120mph being attained whilst serving as a demonstration car at the Paris Motor Show. The gearing was found to be too high, and consequently a modified gearbox, with direct top gear, and rear axle was fitted, the latter having a 3.07:1 ratio. Testing the prototype also showed up problems with its tyres, those normally specified for the standard steel MkVl lasting no more than 20 miles when subjected to prolonged high speeds. Dunlop MDT tyres inflated to 50lbs per square inch were tested, as were India shallow tread tyres, the latter only suitable for speeds up to 115mph. The tyre specified for use with the car was India's Speed Special six-ply, with only half the tread depth of a normal tyre.

The first production Continentals were delivered to Crewe in June 1952. A total of 208 vehicles were built, including the prototype, the last of the series being delivered in May 1955. All but 15 cars were built with Mulliner coachwork, and of those otherwise supplied, two saloons and four drophead coupés were constructed by Park Ward to John Blatchley's styling. Five cars were built in France by Franay, three by Swiss coachbuilder Graber, and one by Pininfarina.

A number of differences existed between the prototype and production cars: for example, the windscreen, instead of being divided, was a full-width single piece, while the roof line was one inch lower; although interior rear headroom was slightly increased. In an effort to keep the weight of the prototype car to an absolute minimum such items as a radio, ashtrays, and even the Bentley mascot, were surrendered. For many customers such sacrifices were unpalatable, and therefore heavier seats, the fitting of a radio along with other accessories increased the payload beyond that originally specified.

Bentley Continental customers found a one-piece slightly curved facia that was different to that fitted to standard steel saloons to give sporting feel. The instrumentation comprised individual gauges rather than the single dial with its four-in-one concept, the rev counter and speedometer being directly ahead of the driver. Because Continental cars were built to customer specification their facia arrangements, while following a general pattern, could be fitted with additional gauges.

The interior reflected the vehicle's sporting attitude: superbly comfortable front bucket seats were usually trimmed with plain leather cushions and pleated backrests. The Continental felt like a sports car, and surprisingly for the size of vehicle the amount of leg room in the front of the cabin was limited. The distance between the bottom edge of the steering wheel and the seat cushion was restricted, the wheel itself being almost vertical and close to the driver's chest. There was also a calmness about the interior, which could accommodate four people, rear seat passengers having more space than first perceived. Luggage space in the sloping tail was limited, but this was a small price to pay for the sheer driving pleasure of this very special touring machine.

Top Right: *When the prototype Continental was retired, Rolls-Royce offered the car for sale. This was unusual because most experimental vehicles were dismantled after undergoing trials. Here the prototype is seen with other Bentleys at a meeting of marque enthusiasts.*

Middle Right: *Bentleys at Le Mans to commemorate the new Bentley Arnage in 1998. Outboard (nearest the camera) are two Continentals, the middle car is an R-type.* Rolls-Royce

Bottom Right: *This Park Ward attired Continental with drophead coupé coachwork is in Scandinavia on official business and contrasts with the surrounding vehicles. Note the Volvo alongside, and a Ford Popular behind.* Rolls-Royce

Top Left: *Looking down at the Bentley stand at the 1952 London Motor Show, the Continental shares space with a standard steel saloon and two coachbuilt cars.* Rolls-Royce

Middle Left: *A year later sees the 1954 models being displayed. The car facing the camera is a Park Ward Continental, that to its left an H.J. Mulliner, and behind is an R-type standard steel saloon.* Rolls-Royce

Bottom Left: *S3 Continentals at the London Motor Show: evident are the styling differences between the Mulliner car (facing) and Park Ward's design created by Vilhelm Koren.* Rolls-Royce

Centre Right: *A derivative of the H.J. Mulliner Bentley Continental was the Flying Spur, a four-door car having much of the two-door car's verve but with improved rear access.*

In 1954 the 4½-litre engine was enlarged to 4887cc and whilst this was initially unique to the Continental, it was fitted to S-series cars the following year. Early Continentals had manual transmission with close-ratio synchromesh gearboxes. Most right-hand-drive-cars had the customary outboard gear selector, although a number were built with central floor change. Left-hand-drive cars were equipped with either central or column change according to customer specification.

Contrary to normal practice, whereby Bentley experimental and prototype cars were destroyed having served their purpose, the prototype Continental was retained as a Bentley demonstrator and was later used as a company car for senior Rolls-Royce and Bentley personnel. Stanley Sedgwick, a long time Bentley enthusiast and president of the Bentley Drivers Club, made it known to Rolls-Royce managers that if ever it was decided to retire the car he would appreciate being offered first refusal. Stanley bought the car in 1960, and although it has changed ownership, the car survives to this day and is a reminder of its outstanding design.

When the S-series superseded the R-type, the Continental's success made it a foregone conclusion that the sporting theme would be continued. Thus the S-type Continental was announced in September 1955. In an initial glance, an S-type Continental with H.J. Mulliner fastback coachwork appears to have similar styling to its predecessor, though a more detailed study of the car reveals significant differences.

Most apparent was the higher straight-through wing line together with rear wings that have pronounced fins.

Out of sight, the chassis with its 4887cc engine shared similar features to that of the saloon. Like its predecessor, the S-series Continental continued to be fitted with lightweight coachwork.

Of the six coachbuilders bodying the S-series Continental, H.J. Mulliner produced 218 cars, Park Ward 185, James Young 20, Hooper six and one each by Franay and Graber. In each instance body styling was unique to the builder, there being the choice of two-door sports saloons in addition to drophead coupés. Facia styling and arrangement were also peculiar to the individual coachbuilder, that of Park Ward models being very different, but no less elegant, to that of its competitors.

Each coachbuilder provided its own touch of distinctive styling, but the most significant departure to Continental styling was that by H.J. Mulliner with four-door coachwork. The car became known as the Flying Spur, but for many buyers it was a more practical option thanks to the increased luggage space, more capacious rear compartment and its additional leg and head room. Conceived as six-light saloons, the Flying Spur was offered as a four-light alternative to afford increased privacy for rear seat passengers. James Young and Hooper also adopted the four-door theme for their models.

When the S2 Continental was introduced in 1959, the previous design was retrospectively known as the S1 model. Apart from having the V8 engine, other chassis changes include a radiator shell with a slightly forward rake and mounted two inches further forward, as well as being three inches lower in height. Coachwork for the S2 was also supplied by H.J. Mulliner who produced 221 vehicles, of which 124 were Flying Spurs; Park Ward built 124 drophead coupés and a single four-door saloon; James Young supplied 41 examples, the majority being four-door saloons, and Hooper a single car which was exhibited at the 1959 Earls Court Motor Show.

An omission to H.J. Mulliner's catalogue was the by now familiar fastback design, though there was some compromise insomuch that the firm's two-door model was a Flying Spur derivative.

A particularly successful offering by Park Ward arrived in the guise of a drophead coupé styled by Vilhelm Koren with John Blatchley's approval. Features of the car were its attractive frontal styling and straight-through wing line which affords a measure of delightful simplicity. Inside the car luxury was maintained in the usual fashion, the facia with its separate driver's instrument nacelle being particularly attractive. The hood, being electro-hydraulically operated, could only be raised or lowered with the gear selector in neutral, and when raised afforded saloon car comfort.

Arrival of the Bentley S3 in 1962 heralded a Continental variant although this time the catalogue did not include a car with a more powerful engine or higher gearing than the standard steel saloon or coachbuilt equivalent. With the acquisition of H.J. Mulliner by Rolls-Royce in 1959, it was amalgamated with Park Ward in 1961 to become Mulliner Park Ward. A total of 291 cars were produced by the combined styling houses and additionally 20 vehicles were built by James Young; a single example was constructed by Graber.

Park Ward's Koren-designed S3 Continental was joined by a stylish fixed-head coupé featuring paired headlamps to good effect, the car's interior being luxuriously equipped and featuring an arrangement similar to the facia of the S2 drophead coupé. The elegance of H.J. Mulliner's S2 Continental and Flying Spur was adapted to S3 guise, but only 11 examples of the former were built before giving way to Park Ward's Koren design, known at Crewe as the 'Korenental'. Unchanged in design, other than restyling of their frontal appearance, the models by the coachbuilder James Young included both two and four-door cars, only two examples of the former being built.

The S3 Continentals were produced until 1965 when the T-series Bentley was introduced, the final chassis being delivered to Mulliner Park Ward in late-November for customer delivery in January 1966. A total of 1,339 R-type and S-series Continentals were therefore built between 1952 and 1965. They are revered by enthusiasts and never fail to achieve immense interest wherever they travel.

A BREED APART - THE BENTLEY CUSTOMER

There wasn't a typical Bentley customer, other than someone able to afford to purchase, run and maintain that select breed of car with its famous insignia. Popular perceptions of the Bentley customer during the period from the end of the Second World War to the mid 1960s will be different to different people, but there is likely to be at least a couple or so common images, one being of the affluent tweeds attired 'country gent', another the successful businessman, and possibly a sheepskin clad Warren Street motor trader.

Above: *Bentley customer and an ex-Chairman of Bentley Motors, Captain Woolf Barnato with his 1936 4¼-litre Derby Bentley with Park Ward coachwork . When Barnato was unable to maintain his investment in Bentley Motors in 1931, the firm was taken over by Rolls-Royce, after which he became a director of the newly formed Bentley Motors (1931) Limited.*

A list of Bentley customers might have included racing drivers, stars of stage and screen, bankers, directors, doctors, lawyers, landowners, nightclub owners, property developers, sports promoters, aristocracy, heads of state and royalty.

And then there was the loyal Bentley enthusiasts for whom no other car would suffice. Motor sport personalities associated with the Bentley marque are numerous, including the late Rivers Fletcher who apprenticed with Bentley Motors when WO was in charge, S.C.H. 'Sammy' Davis who was one of the Bentley Boys, as was Jack Dunfee. Tony Vandervell was an advocate of the Bentley Continental, and for Raymond Mays of ERA and BRM fame his MkVI was the first of several post-war Bentleys. During pre-war days motoring 'aces' Sir Malcolm Campbell and Captain Eyston both owned Bentleys.

A glance at the list of R-type Continental owners reveals some interesting names: American racing driver Briggs Cunningham, millionaire Aristotle Onassis, Nicholas Monsarrat, author of *The Cruel Sea*; property tycoon Sir Alfred McAlpine and clothing magnate R. Montague Burton.

Other personalities associated with the marque include Henry Ford II who ordered an R-type two-door saloon; Leonard Lord of the British Motor Corporation; Sir John Black, head of Standard-Triumph, and actor Stewart Granger whose MkVI fixed head coupé had Freestone & Webb coachwork. Bentleys were purchased for use by company Chairmen and senior executives for well known firms like Fox's Glacier Mints Ltd., Godfrey Davis Ltd., Vernon's Mail Order Stores, Turner & Newall Ltd., British Aluminium Co. Ltd. and Lilley & Skinner.

Bentleys, along with other expensive makes of car, tended to change hands at more frequent intervals than many other vehicles. There were reasons for this as such cars held and often increased their values, therefore making the purchase of subsequent models all the more attractive. There were those customers who perceived that to be seen buying the latest Bentley was a sign of continuing wealth and success.

Top Right: *Sir Malcolm Campbell, the legendary world speed record-holder noted for his* Bluebird *speed vehicle, preferred something more comfortable as his everyday mode of transport; here he is with his Derby Bentley, the forerunner to the Crewe models.*

Middle Right: *Raymond Mays of ERA and BRM fame is seen here with his Derby Bentley. When the MkVI appeared in 1946, Mays became a strong advocate for the car.*

Bottom Right: *World speed record holder Captain George Eyston with his Derby Bentley. It was Captain Eyston who put the Pourtout-bodied Bentley through its paces at Brooklands in 1939, that car inspiring the Continental's development.*

Above: *A scene pictured within Rolls-Royce's London Service Centre at Hythe Road in Willesden. A variety of cars can be seen, that nearest the camera a MkVl. Removal of the passenger seat cushion has exposed the car's heater. Removal of the driver's cushion would reveal the car's battery and brake master cylinder.* Rolls-Royce

Purchasing a new Bentley was a satisfying experience. A mere glance at a Bentley brochure instilled the feeling of well-being, especially as the car commanded one of the highest values in the tables printed in *The Motor* and *The Autocar*. There was a choice when it came to the method of buying one's new Bentley: when customers visited Rolls-Royce and Bentley's London showroom at Conduit Street they were greeted by Jack Scott, Head of Sales.

There were two telephones on Jack's desk; on the ringing of one he answered 'Bentley Motors', the other 'Rolls-Royce Limited. However, other customers preferred to do business with the well known London Bentley and Rolls-Royce dealers, such as Jack Barclay of Berkeley Square, Jack Olding & Co. of North Audley Street W1, Charles Follett of Mayfair, H.R.Owen of Berkeley Street or Car Mart of London's Park Lane, the latter handling sales for the Royal Household.

In the provinces, dealers like Weybridge Automobiles, Edwards & Co of Bournemouth, Caffyns, Rippon Bros. of Huddersfield, and Loxhams of Preston and Blackpool offered sales, servicing and coachwork repairs to the highest quality. Indeed, Rippon's used to say that per head of population Huddersfield had more Rolls and Bentley models than any other town or city in the world.

For those customers residing in more remote regions of the British Isles, Bentley sales representatives such as David Buckle and David Tod went to them, both responsible for Bentley and Rolls-Royce sales throughout the whole of Scotland at different times. Some Bentley dealers had motor sport connections. Mike Couper, for example, had a Rolls-Royce and Bentley agency at St Albans in Hertfordshire, and regularly participated in the Monte Carlo Rally driving the latest Bentley. Couper's experience, and that of his co-drivers, meant that on more than one occasion not only was he a class winner but holder of the Grand Prix in the Concours de Confort.

Rallying wasn't like it is today: then it was a thoroughly gentlemanly affair when it was customary to wear either a suit, or at least a sports jacket, whilst at the wheel, and routine lunch and dinner stops were made at the finest hotels en route.

Ordering the latest Bentley MkVl, R-type or S-series was a painstaking business, even when purchasing a standard steel saloon.

Selecting a coachbuilt car called for all the more thoroughness, the customer being directed to the coachbuilder of their choice. Respected coachbuilders the like of H.J.Mulliner being quite different to Freestone & Webb, Hooper and James Young were recognised for their individual designs and unique interior refinements. The customer was often invited to the coachbuilder's premises where he or she would discuss a range of details, from a particular coachwork design, along with any individual modifications or enhancements, to specific choice of interior furnishings and accessories. From the time a chassis arrived at the coachbuilder's premises it could take several months before the car was delivered.

Below: *Another scene showing Bentleys in various stages of maintenance and repair at Hythe Road, the MkVI in the foreground having a front end overhaul. Note the car next to it on axle stands, the sign of a costly undertaking!*
Rolls-Royce

Above: *The elegance of this James Young-bodied R-type saloon pictured in London's Kensington Gardens on a summery day is immediately apparent. The design of the coachwork anticipates to some extent the styling adopted by John Blatchley for the S-series Bentley.*

Left: *Bentley cars were often the chosen transport of professional people, and as such more than a prominent status symbol. However, they could also be hard-working cars too, and some of the locations in which they have been filmed are hardly glamorous. In this superb image we see a Cheshire registered MkVI or R-type pictured in wintry conditions amidst remote scenery.* Rolls-Royce

Coachbuilder Harold Radford, for example built a number of Bentleys with the hunting, fishing and shooting customer very much in mind. His first foray into Bentley conversion was after World War II (in 1948) when he displayed a MkVI chassis that he had bodied with wooden panelling in the style of a shooting brake. at the Earls Court Motor Show where it was dubbed The Countryman.

The car was fitted out with every conceivable accessory to attract a satisfactory number of orders. When it became too expensive to build the Countryman from scratch, the coachbuilder acquired standard steel saloons from Crewe and fitted them out to customer specification. Harold Radford's Countryman conversions on the S-series Bentley were particularly successful, each car being unique to customer requirement.

Above: *Here we have a picture showing Bentley Drivers Club members meeting at Crewe with their cars in the early post-war years. It is clear from this image that the MkVIs are outnumbered by 'WO' models. Of course, Bentleys have always attracted enthusiastic motorists who have become loyal to the marque.*
Rolls-Royce

The Radford theme was extended to a Bentley Town & Country Estate Car conversion of the S-series, the external modifications being undertaken by H.J. Mulliner. Harold Radford carried out the interior specification, which included provision of folding rear seats that formed a flat floor aft of the front seats to the split tailgate.

In addition to there being a nationwide network of dealers, Rolls-Royce and Bentley operated a London service centre at Hythe Road in Willesden. A hive of industry, this service centre attended to every aspect of Rolls-Royce and Bentley ownership, from routine maintenance to complete overhauls, bodywork repairs and full restoration of older cars. Here, as elsewhere in the organisation, attention to detail and high levels of customer service were the order of the day and these in turn made for satisfied patrons.

Of course, not all customers bought brand new Bentleys, as there was also a very good market in used models. Generally most used examples were still in an immaculate condition when they were being traded in, and as such they were well sought-after.

People with an eye for a bargain were attracted to the marque when looking for a used luxury car. In the mid-1950s a late-1940s MkVI standard steel saloon could be purchased for less than £1,000 although such value gave little indication as to the car's condition. Coachbuilt examples commanded higher values according to the particular coachbuilder and style of coachwork. By comparison a similar age Jaguar could be acquired for under £400.

For a lot of owners the cost of properly maintaining a Bentley proved to be vastly more expensive than they had anticipated and quickly sold their cars on or allowed them to depreciate to a poor condition. By the mid-1960s values of early post-war Bentleys, especially standard steel saloons, had diminished to such a point that they were often virtually worthless, and even cherished vehicles were fetching prices below £300-£400. My editor recalls one being used by a plant-hire firm owner to take him to hospital after being attacked by a dog. Despite the editor's arm bleeding profusely all over the rear seats in two-tone leather, the owner was quite un-phased by the staining saying, that his car was worth less than one of his ex-US Army Jeeps.

Today, post-war Bentleys built before 1964 are finding enthusiasts who are appreciative of the marque's heritage and quality, and keen to maintain them in the style originally intended. Values mostly mirror a model's exclusivity, some of the most sought after Bentleys being coachbuilt examples, particularly the R-type Continental.

Top Right: *The interior of Rolls-Royce and Bentley's London Conduit Street showroom. Nearest the camera is a MkVI with Park Ward coachwork, a standard steel saloon is visible in the background. Adjacent to the window is a Rolls-Royce Silver Wraith, and outside, a Bentley demonstrator.*
Rolls-Royce

Middle Right: *Nowadays, the superb luxury and elegance of post-war Bentleys make for ideal wedding vehicles, as they have both a unique presence combined with spaciousness and a feeling of quality that is simply not seen in the modern motor cars of today.*

Bottom Right: *A MkVI standard steel saloon leading a coachbuilt drophead coupé at a Rolls-Royce and Bentley enthusiasts' rally at Harewood House in Yorkshire. Cars like these command a loyal following by enthusiasts and a substantial price tag for those who would join them.*

Above: *Customers who entrusted their Bentleys to Rolls-Royce's London Service Centre were assured the best attention. In this instance the service representative is discussing the schedule of work with the vehicle owner, repairs that could well include rectifying minor accident damage to the front offside wing.* Rolls-Royce

Right: *Few cars enjoy the elegance of this H.J. Mulliner bodied MkVI pictured in London amidst delectable surroundings. The car was delivered to Iliffe & Sons Ltd. in July 1952. H.J. Mulliner was acquired by Rolls-Royce in 1959 and merged with Park Ward to become H.J. Mulliner, Park Ward, and later simply Mulliner Park Ward.*

Above: *This impressive gathering of Bentleys at Crewe depicts mainly early post-war models, though a few 'WO' cars are evident. Two organisations cater for the Bentley enthusiast, the Bentley Drivers Club and the Rolls-Royce Enthusiasts' Club. The former is for all Bentley devotees, the latter for enthusiasts of Derby- and Crewe-built cars.*

ACKNOWLEDGEMENTS

No book of this kind could be possible without the kind assistance of a large body of people so the author and publishers wish to thank all those who have made this publication possible. In particular John Blatchley who discussed at length the cars he created; Martin Bourne, a retired senior stylist at Crewe who worked with John Blatchley and who recalls with pride and enthusiasm his career with Rolls-Royce; Richard Mann, whose Rolls-Royce and Bentley years were spent at the company's London coachwork facility as a quality engineer; and not least to motoring historian and researcher Andrew Minney for his valuable assistance.

Unless otherwise indicated the illustrations are from the author's collection. The author acknowledges assistance provided by Rolls-Royce Motor Cars Limited, now Bentley Motors, in providing a number of images. Grateful thanks are extended to LAT for use of archive material, and to Bentley owners who allowed their cars to be photographed.